# My Space Book

## By Nicholas J. D. Sims

Copyright © 2023 by Nicholas J.D. Sims

All rights reserved. This book or any portion thereof may not be reproduced or used in any manner whatsoever without the express written permission of the publisher except for the use of brief quotations in a book review.

Once upon a time, in a far-off corner of the universe, there was a brilliant star named the Sun.

The Sun was the center of a great and wonderful family, made up of eight planets, each with their own special characteristics.

# MERCURY

The first planet closest to the Sun was tiny Mercury.

Mercury was so small that it only took 88 days to orbit the Sun, it was the fastest of all the planets.

# VENUS

**Next came Venus, the brightest planet in the sky. Venus was covered in thick clouds, and it was so hot that it could melt lead.**

# EARTH/MOON

**Earth was the third planet from the Sun. Earth was the only planet that had life, and it was where we lived. It took 365 days for Earth to orbit the Sun.**

Mars was the fourth planet from the Sun. Mars was known as the "Red Planet" because of its reddish color. It was also home to the largest volcano and canyon in the solar system named Olympus Mons.

# JUPITER

Jupiter was the fifth planet from the Sun. Jupiter was the biggest planet in the solar system, and it was so big that all the other planets could fit inside of it. It had a giant storm called the "Great Red Spot" that was bigger than Earth.

# SATURN

Saturn was the sixth planet from the Sun.

Saturn was known for its beautiful rings, made up of ice and dust.

# URANUS

**Uranus was the seventh planet from the Sun. Uranus was the only planet that orbited the Sun on its side, making it look like it was rolling through space.**

# NEPTUNE

**Lastly, was Neptune, the eighth planet from the Sun. Neptune was the farthest planet from the Sun and it was so cold that it had frozen methane on its surface.**

These eight planets, along with many other celestial bodies, formed the solar system, a place of wonder and beauty that we are still exploring today.

The end.

# I HAVEN'T FORGOTTEN ABOUT PLUTO

For years, Pluto has been a little planet that just didn't quite fit in with the rest of the solar system. That's because it's a dwarf planet—a celestial body that is not a full-fledged planet like Mercury, Venus, Mars, Jupiter and Saturn.

Instead, Pluto is technically a dwarf planet "a large enough body whose orbital path lies entirely in or beyond the plane of the planets." This means that Pluto still has lots of asteroids and other space rocks along its flight path, rather than having absorbed them over time, like the larger planets have done.

www.ingramcontent.com/pod-product-compliance
Lightning Source LLC
Chambersburg PA
CBHW042251100526
44587CB00002B/107